Real sequences and limits

A collection of solved problems

Ekrem Hamidovic

2

Foreword

This collection of problems is aimed to give a guidance to student taking Calculus I and having a major in mathematics, engineering, physics or any related field and introduce to the methods and concepts of real sequences and their features.

As a first part we give an overview of basic concepts and theorems (stated without proof) and then proceed to solved problems. Even though mathematics can't be understood without proper practice, we think that a student who is new to the subject should see the demonstrated solutions with comments. Of course, student is always encouraged to try to solve without referring immediately to the solution. And to bring another solution as well.

I hope that anyone who uses the book will be able to see the beauty of the methods of calculus, and to realize the importance of the concepts later in the broader context. The examples presented here have been chosen so to show the variety of methods that are applied in theory of sequences. It is far from any exhaustive list of such methods. Student is highly encouraged to continue her own research in case of further interest.

Author

Sequences - concepts and theorems

1.1 Key words

The following are the key words in this section: sequence, subsequence, increasing and decreasing sequences, monotonic sequences, bounded sequences, infimum, supremum, limit superior, limit inferior, convergence, divergence.

1.2 Definitions and theorems

1.2.1 Definitions

Sequence, subsequence, convergence

Definition 1.2.1. *A sequence of real numbers is any mapping* $f : \mathbb{N} \to \mathbb{R}$. *Sequence are denoted by* $\{a_n\}$ *to replace the form* $a_1, a_2, ..., a_n, ...$ *where* $f(i) = a_i$.

Definition 1.2.2. *Let* $\{a_i\}_{\mathbb{N}}$ *be a real sequence. Let* $J \subset \mathbb{N}$ *be any strictly increasing sequence of natural numbers. Then* $\{a_j\}_{j \in J}$ *is called the subsequence of the sequence* $\{a_i\}_{\mathbb{N}}$.

Definition 1.2.3. *A sequence of real numbers is said to be convergent to* a, *if for every positive number* ε *there exists a natural number* n *depending* *on ε so that for all $m > n$ the inequality*

$$a_m - a < \varepsilon$$

holds. We write $\lim_{n \to \infty} a_n = a$ and denote a as the limit of the sequence a_n.

monotonic sequences, bounded sequences

Definition 1.2.4. *A sequence is said to be*

1. ***increasing*** *if for any two consecutive members of the sequence a_i, a_{i+1} the inequality $a_i \leq a_{i+1}$*

2. ***strictly increasing*** *if for any two consecutive members of the sequence a_i, a_{i+1} the inequality $a_i < a_{i+1}$*

3. ***decreasing*** *if for any two consecutive members of the sequence a_i, a_{i+1} the inequality $a_i \geq a_{i+1}$*

4. ***strictly decreasing*** *if for any two consecutive members of the sequence a_i, a_{i+1} the inequality $a_i > a_{i+1}$*

5. ***bounded from above*** *if there exists a real number M such that for all, except possibly finitely many, members of the sequence $a_n \leq M$.*

6. ***bounded from below*** *if there exists a real number m such that for all, except possibly finitely many, members of the sequence $a_n \geq m$*

7. ***bounded*** *if it is bounded from above and below.*

Supremum and infimum

1. **(Supremum)** Let $\{a_n\}$ be a real sequence bounded from above by M. M is said to be the lowest upper bound or supremum of $\{a_n\}$ if the following condition holds:

$$\text{for all upper bounds of } \{a_n\} : M \leq M' \tag{1.2.1.1}$$

2. **(Infimum)** Let $\{a_n\}$ be a real sequence bounded from below by m. m is said to be the greatest upper bound or infimum of $\{a_n\}$ if the following condition holds:

$$\text{for all lower bounds of } \{a_n\} : m \geq m' \qquad (1.2.1.2)$$

1.2.2 Theorems

Convergence and limits theorems

1. The limit of a sequence is unique.

2. Any convergent sequence is bounded. Hence, an unbounded sequence is not convergent.

3. Any subsequence of a convergent sequence is convergent towards the same number.

4. If a finite number of members of a sequence is changed or deleted, that doesn't change the convergence of the sequence or its limit, if it exists.

5. A sequence is divergent if it has at least one divergent subsequence.

6. If $a_n \to 0$ and b_n is bounded sequence, then $a_n b_n \to 0$.

Operations on limits

Let $a_n \to a, b_n \to b$ be real convergent sequences and c a constant.

1. $a_n \pm b_n \to a \pm b$.

2. $a_n \cdot b_n \to ab$.

3. Provided $b_n \neq 0, b \neq 0$, $\dfrac{a_n}{b_n} \to \dfrac{a}{b}$.

4. $c a_n \to c a$.

Convergence criteria

1. **(Monotony)** A monotonic sequence converges only if it is bounded. If it is decreasing, then it converges to the infimum; if it is increasing, it converges to the supremum.

2. **(Cauchy)** A sequence $\{a_n\}$ converges only if it satisfies the Cauchy-condition of the convergence: for all $\varepsilon > 0$ there exists $N \in \mathbb{N}$ such that for all $m, n > N$:
$$| a_m - a_n | < \varepsilon$$

3. **(Bolzano-Weierstrass)** Every bounded sequence, convergent or divergent, contains at least one convergent subsequence.

4. **(Nested intervals theorem)** If $\{I_n\}$ is a sequence of intervals such that $I_1 \supset I_2 \supset I_3 \supset \cdots$ with $I_n = [a_n, b_n], n \in \mathbb{N}$, then there exists a unique real number contained in the intersection of the whole sequence: $I_1 \cap I_2 \cap \cdots = \{a^*\}$.

Consequences and further theorems

1. (The supremum principle) Every bounded sequence of real numbers has the least upper bound and the greatest lower bound in $\mathbb{R}$.

2. Every Cauchy sequence is bounded.

3. Every Cauchy sequence is convergent.

4. Every convergent sequence is also a Cauchy sequence.

Problems and solutions

Question 1. Find the limit of the sequence $\frac{1}{n}, n \to \infty$

Solution. This problem can be solved in several ways. One obvious way is to look at the members of the sequence. Notice that the sequence is bounded from above by $a_1 = \frac{1}{1}$ and that the sequence is decreasing: $n_1 > n_2 \Rightarrow \frac{1}{n_1} < \frac{1}{n_1}$. Furthermore we can see that all the members of the sequence are positive and as $n \to \infty$ the sequence is getting closer to 0. This indicates that 0 is possible candidate for the limit. Indeed, for every $\varepsilon > 0$ there is $N \in \mathbb{N}$ such that for all $n > N : \frac{1}{n} - 0 = \frac{1}{n} < \varepsilon$. This comes from the Archimedean property of real numbers or from the fact that the set of natural numbers is not bounded from above. Following either of these arguments the desired inequality can be easily established - for every $\varepsilon > 0$ let $N > \frac{1}{\varepsilon}$. Then the required inequality follows. So the limit of the sequence is indeed zero.

Question 2.

Analyze the convergence of the sequences of the form

$$a_n = \frac{1}{n^\alpha}$$

where $\alpha \in \mathbb{Q}$.

Solution. The following three cases are to be considered:

9

1. $\alpha \geq 1$

2. $0 < \alpha < 1$

3. $\alpha < 0$

- If $\alpha < 0$ then this sequence has the form $n^\beta, \beta > 0$, which is not bounded. Hence it is not convergent.

- If $\alpha \geq 1$, the sequence is convergent and in fact in converges to 0. The inequality:

$$n^\alpha \geq n$$

for all $n \in \mathbb{N}$ and for such α is obvious, so that

$$\frac{1}{n^\alpha} < \frac{1}{n}$$

and applying the fact that $\frac{1}{n} \to 0$ the result follows.

- If $0 < \alpha < 1$, then $\alpha = \frac{1}{m}$ for some $m \in \mathbb{N}$. So that the sequence has the form $\frac{1}{\sqrt[m]{n}}$. This sequence converges towards 0. The inequality:

$$\sqrt[m]{n} < n$$

for all $n \geq 1$, $\sqrt[m]{n} < n$. This can be proven by two inductions: by fixing m prove on induction over n that the inequality holds. Then suppose the induction hypothesis for m and prove again by induction over n that it holds for $m + 1$. The essence is the fact that for any $m, n \in \mathbb{N}$ we have:

$$n \leq n^m$$

an inequality which also can easily be established by a straightforward induction.

Question 3. If $a_n \to c_0, b_n \to d_0$, to where does the sequences

$$\sum_{k=0}^{m} b_k \{a_n\}^k \qquad \frac{\sum_{k=0}^{n} h_k \{a_n\}^k}{\sum_{k=0}^{m} s_k \{b_n\}^k}$$

converge?

Solution. Such a sequence

$$b_0 + b_1\{a_n\} + b_2\{a_n^2\} + \cdot + b_m\{a_n^m\}$$

is indeed convergent, as the finite sum and finite product of a convergent sequence are convergent. The limit is simply

$$\sum_{k=0}^{m} b_k c_0^k$$

As for the second sequence, it converges only if $m > n$. The limit is $\dfrac{h_0}{s_0}$.

Question 4. Show that there is no sequence that contains all real numbers.

Solution. This statement is a sequential analogy of the fact that there is no surjective mapping between $\mathbb{N}$ and $\mathbb{R}$.

Question 5. Prove the following:

1. If $x_n \to a$, then $|x_n| \to |a|, n \to \infty$.

2. If $\{x_n\}$ is monotone and contains a convergent subsequence, then it is convergent to the same limit as its subsequence.

Solution.

1. The proof is based on the inequality $||x_n| - |a|| < |x_n - a| < \varepsilon$, since for any two real numbers $x, y : ||x| - |y|| < |x - y|$ and the last inequality follows from the assumption that $x_n \to a$.

2. Suppose that the sequence x_n is increasing (if it's decreasing, just apply the argument to the sequence $-x_n$). Let y_n be it's convergent

subsequence (note that y_n is increasing as well). By assumption: for all $n > N(\varepsilon) : |y_n - a| < \varepsilon$ and hence $y_n < \varepsilon + a$. On the other side we have

$$x_n \leq x_{n+1}$$

for all but finitely many n and especially we can say for all $n > N(\varepsilon)$ this holds. If that is not the case, choose the smallest M for which the following conditions hold:

(a) for all $n > M : y_n < a + \varepsilon$

(b) for all $n > M : x_n < x_{n+1}$

Both conditions are fulfilled based on the assumptions. How are the subsequence y_n and the sequence x_n related except that the later is the subsequence of the prior? The gap between the members of the sequence y_n such that there is no member of x_n can be at most finite. The same applies for the sequence x_n - there can be only finite gap between members of y_n. That is:

$$x_n \leq \underbrace{\quad \cdots \quad}_{\text{no elements of } y_n} \leq x_{l+n} \leq \underbrace{\quad \cdots \quad}_{\text{no elements of } x_n} \leq y_k \leq \cdots$$

Neglecting all such (finitely many) gaps, we can see that the whole sequence x_n is dominated by some subsequence on y_n call it z_n so that for all but finite many $n : x_n < z_n$ and hence for all but finite many $n : x_n < a + \varepsilon$. This shows the convergence to the same limit a, since x_n is increasing and its supremum is a.

Question 6. Give an example of a sequence which is bounded, but isn't convergent.

Solution. The sequence defined as: $a_{2k} = 1, a_{2k+1} = -1$ is an example of the required sequence.

Question 7.   Calculate the limit:

$$\lim_{n\to\infty} \sqrt{\frac{n^2+1}{n+3}} + n$$

Solution.

$$\lim_{n\to\infty} \sqrt{\frac{n^2+1}{n+3}} + n = \lim_{n\to\infty} \left(\sqrt{\frac{n^2+1}{n+3}} + n\right) \frac{\sqrt{\dfrac{n^2+1}{n+3}} - n}{\sqrt{\dfrac{n^2+1}{n+3}} - n}$$

$$= \lim_{n\to\infty} \frac{\dfrac{n^2+1}{n+3} - n^2}{\sqrt{\dfrac{n^2+1}{n+3}} - n} = \lim_{n\to\infty} \frac{n^3+1-n^3-3n^2}{(n+3)\sqrt{\dfrac{n^3+1}{n+3}} - n(n+3)}$$

$$= \lim_{n\to\infty} \frac{-3+\dfrac{1}{n^2}}{-\left(1+\dfrac{3}{n}\right)\sqrt{\dfrac{n+\dfrac{1}{n^2}}{n+3}} - 1 - \dfrac{3}{n}} = \frac{3}{2}$$

Question 8.   Show that $\lim\limits_{n\to\infty} \dfrac{n}{2^n} = 0.$

Solution.   We will show that for every $\varepsilon > 0$ and all $n > N$, where $N > 1 + \dfrac{2}{\varepsilon}$ we have $\dfrac{n}{2^n} < \varepsilon$.

$$\frac{n}{2^n} = \frac{n}{(1+1)^n} = \frac{n}{1+n+\dfrac{n(n-1)}{2}+\cdots+1} < \frac{n}{\dfrac{n(n-1)}{2}} = \frac{2}{n-1}$$

So if we want $\dfrac{2}{n-1} < \varepsilon$, then solving inequality we obtain: $n > 1 + \dfrac{2}{\varepsilon}$. So for every $N > 1 + \dfrac{2}{\varepsilon}$ the inequality holds.

Question 9.   Prove that $\lim\limits_{n\to\infty} na^n = 0, |a| < 1.$

Solution. When $a = 0$, the sequence reduces to the constant sequence of zeros. Assume $a \neq 0$. Since $\mid a \mid < 1$, it follows: $\left| \dfrac{1}{a} \right| > 1$. Therefore, we can set $\left| \dfrac{1}{a} \right| = 1 + c, c > 0$ or $|a| = \dfrac{1}{1+c}$. Assume that $n > 2$. Then:

$$|na^n| = \frac{n}{(1+c)^n} = \frac{n}{1 + nc + \dfrac{n(n-1)}{2}c^2 + \cdots + c^n} < \frac{n}{\dfrac{n(n-1)c^2}{2}} = \frac{2}{(n-1)c^2}$$

As the last sequence tends to zero, the statement follows.

Question 10. Prove that for $\alpha > 0, |a| < 1 : \lim\limits_{n \to \infty} n^{\alpha}a^n = 0$.

Solution. This follows from the previous question with some transformations. If we set:

$$n^{\alpha}a^n = n^{\alpha}a^{\alpha/\alpha n} = n^{\alpha}\left(a^{1/\alpha n}\right)^{\alpha} = (nb^n)^{\alpha}$$

with $b = a^{1/\alpha}, \alpha > 0$. Since $\alpha > 0$ and $|a| < 1$, it follows that $|b| < 1$. Now apply the previous question on the sequence nb^n - it tends to zero. Hence $(nb^n)^{\alpha} \to 0, n \to \infty$.

Question 11. Prove that $\dfrac{\ln}{n^{\alpha}}$ converges to zero for $\alpha > 0$.

Solution. This question can be solved in different ways.

First method With a transformation:

$$\frac{\ln n}{n^{\alpha}} = \frac{\ln n^{\alpha/\alpha}}{n^{\alpha}} = \frac{\ln (n^{\alpha})^{1/\alpha}}{n^{\alpha}} = \frac{\ln n^{\alpha}}{\alpha n^{\alpha}}$$

the problem is reduced to showing

$$\frac{\ln n}{n} \to 0, n \to \infty$$

We notice that $1 \le n \le e^n, n \in \mathbb{N}$ so that $\ln n \le n \in \mathbb{N}$ and hence:

$$0 \le \frac{\ln n}{n} \le 1$$

is bounded. Also for $m > n$ we have:

$$\frac{\ln n}{n} - \frac{\ln m}{m} < \frac{\ln n}{n} - \frac{\ln m}{n} = \frac{\ln n - \ln m}{n} < 0$$

showing that $\left(\frac{\ln n}{n}\right)$ is a decreasing sequence, which converges to its minimum.

Second method Using the completeness axiom of real numbers or any equivalence of it, there is for every $n \geq 2 \in \mathbb{N}$ a natural number c_n such that:

$$e^{c_n} \leq n \leq e^{c_n+1}$$

so taking natural logarithm of both sides we obtain:

$$\frac{\ln n}{n^\alpha} \leq \frac{c_n + 1}{e^{\alpha c_n}} = e^\alpha \frac{c_n + 1}{(e^\alpha)^{c_n+1}}$$

Taking $a = \frac{1}{e^\alpha}, c_n + 1 = n$, we obtain

$$\frac{\ln n}{n^\alpha} < e^\alpha n a^n$$

Now referring to the fact that $n a^n \to 0, |a| < 1$, the result follows.

Third method We know that

$$\frac{n}{b^n} \to 0, n \to \infty$$

In terms of bounds, for n large enough:

$$\frac{n}{b^n} < 1$$

and also obvious inequality:

$$\frac{1}{b^n} \leq \frac{n}{b^n}$$

Let $b = a\varepsilon, a > 1, \varepsilon > 0$. Then we have:

$$\frac{1}{a^{\varepsilon n}} < \frac{n}{a^{\varepsilon n}} < 1$$

that is

$$1 < n < a^{\varepsilon n}$$

Taking logarithms of both sides, we obtain:

$$0 < \ln n < n\varepsilon \ln a$$

or

$$0 < \frac{\ln n}{n} < \varepsilon \ln a < \varepsilon$$

Question 12. Prove that $\dfrac{a^n}{n^n} \to 0, n \to \infty, a \in \mathbb{R}$.

Solution.

First method   First we will consider the case $a \in \mathbb{N}$. The proof is based on the inequality:

$$a^n < n^n$$

which is proven by an induction with the smallest n such that $n > a$. So for every $a \in \mathbb{N}$ the induction basis will start at $n = a+1$. Let us fix $a \in \mathbb{N}$. For the induction step we have

$$a^{n+1} = a^n a < n^n \cdot a < n^n \cdot n < (n+1)^{n+1}$$

This shows that the sequence is bounded from above by 1 and it is also bounded from below by 0. Furthermore we have:

$$\frac{x_{n+1}}{x_n} = \frac{\dfrac{a^{n+1}}{(n+1)^{n+1}}}{\dfrac{a^n}{n^n}} = \frac{an^n}{(n+1)^{n+1}} \leq 1$$

which shows that the sequence is decreasing, taken however from sufficiently large n. This is consistent with the argument, as we are interested in a limiting process, hence leaving first a members doesn't affect the convergence. Hence the convergence. However, we must show that it indeed converges to zero. Here we can use the following argument: when a sequence is known to be convergent, then the limit of any subsequence will be also the limit of the original sequence. If we take the subsequence of the form

$$\frac{a^n}{(ma)^n}$$

then we can see this subsequence converges to zero. Hence, the original sequence converges to zero as well. What we have proven is:

$$\forall a \in \mathbb{N} : \frac{a^n}{n^n} \to 0, n \to \infty$$

The proof for $a \in \mathbb{R}$ is now carried out by majoring any given $\left|\frac{a^n}{n^n}\right|$ by a suitably chosen $b \in \mathbb{N}$ and sequence $\frac{b^n}{n^n}$.

Second method In analogy to the first method, instead of $\frac{a^n}{n^n}$ we can take any b larger than a. Applying the proven inequality $a^n < n^n$ to b, we can simply estimate: $\left(\frac{a}{n}\right)^n < \left(\frac{a}{b}\right)^n$ for sufficiently large n with the condition $a < b$. The later sequence is known to be convergent, as $\left|\frac{a}{b}\right| < 1$.

Question 13. Prove the following: Let $\{a_n\}$ be a sequence with $a_n \to a$. Then for the sequence of arithmetic means of the sequence a_n it holds:

$$\frac{a_1 + \cdots + a_n}{n} \to a$$

as well as the sequence of the geometric means

$$\sqrt[n]{a_1 \cdots a_n}$$

Solution. First we prove a special case when $a = 0$. If $a_n \to 0$, then for $n > m$:

$$|a_n| < \frac{\varepsilon}{2}$$

From here:

$$\left| \frac{a_1 + \cdots + a_n}{n} \right| \leq \frac{|a_1 + \cdots + a_m|}{n+1} + \frac{\varepsilon}{2} < \frac{\varepsilon}{2} + \frac{\varepsilon}{2} = \varepsilon$$

The first summand is fixed number and for large n it can be approximated to be less than $\varepsilon/2$.

Regarding the second part, it is a simple consequence of the inequality

$$\sqrt[n]{a_1 \cdots a_n} \leq \frac{a_1 + \cdots + a_n}{n}$$

taking into consideration the domain where the expressions are defined.

Question 14.

1. (Monotonic Subsequence Theorem) Every sequence of real numbers contains at least one monotonic subsequence.

2. Every increasing sequence bounded above converges to the supremum. Every sequence bounded from below converges to the infimum.

3. (Bolzano-Weierstrass) Every bounded sequence has a convergent subsequence.

Solution.

1. According to the law of trichotomy of real numbers, any two real numbers can be compared.

   Let us call an element c_m of a sequence *upper local vertex* if there exist an indexing of the sequence $\{a_n\}$ such that infinitely many members of that sequence are less than c_m. Similarly, a sequence has a *lower local vertex* b_m if there is an indexing of the sequence such that infinitely many members are greater than b_m.

   Such an indexing is most general, and doesn't need to follow any formal pattern. Furthermore a gap between any two members in such an indexing can be arbitrarily large - even denumerably large. Local vertex can exist, there can be more than one local vertex, there can be infinitely many local vertices but may also not exist.

(a) If a sequence is constant or monotonic (increasing or decrasing, non-increasing or non-decreasing), it contains infinitely many local vertices.

(b) Also, a sequence with infinitely many local vertices is either constant or monotonic.

(c) In a constant sequence, each element is both upper and lower local vertex.

(d) A sequence which splits into p constant subsequences has p local vertices.

(e) Any local vertex induces a tree within the original sequence with finitely many branches of which at least one forms an infinite monotonic sequence. The proof is constructive and it applies in each stage the law of trichotomy. Let A be an upper local vertex. Then A induces a set consisting of members of the original sequence with the corresponding index so that $A \geq a_k, k \in \{i_1, i_2, ..., i_n, ...\}$. Denote that set with B_n. Looking at the set $A \cup B_n$, we see that A is its maximum. However B_n itself has the supremum, since it is bounded from above by A. The total ordering on B_n implies that B_n can be partitioned into at most three possibly distinct linearly ordered sets:

  i. those containing copies of one element

  ii. those containing distinct elements which are

   A. either decreasing

   B. or decreasing

The final step is to notice that at least one of these partitioning must be infinite, for otherwise, if all are finite, the B_n would be finite too, which is a contradiction. Now choose from this infinite sequence an infinite subsequence indexing it by a suitable linear function. This subsequence is the required one. In case A is a lower local vertex, the proof is analoguous.

(f) There are only two types of sequences: those having a finite (including zero) and those with infinite many local vertices.

From (f) we have to consider two cases: sequences with a finite and infinite many local vertices. Based on (a) and (b) cases for zero and infinite numbers of local vertices. Statement (e) is examination of the case for finite many local vertices.

2. Let $\{a_n\}$ be an increasing sequence and bounded from above. According to the supremum principle, there exists $\sup a_n = A$. For every $\varepsilon > 0$ we have $|a_n - A| < \varepsilon$ for sufficiently large n. Indeed, we know that for any $\varepsilon > 0, A - \varepsilon$ is not even an upper bound. This comes from the definition of the supremum. Hence, for some $m \in \mathbb{N}$ we have that $a_n > A - \varepsilon$ for all $n > m$. Also $a_n < A + \varepsilon$ for $n > m$. Together this shows $|a_n - \varepsilon| < A$.

3. The theorem follows from 1. and 2.

Remark. Part 1 of the previous question can be solved using Ramsey's theorem. The proof is based on the combinatorial interpretation of the law of trichotomy. For any pair (a_i, a_j) of the sequence color it

- blue, if $i = j$

- red, if $i > j$

- green, if $i < j$

Ramsey's theorem implies that there is infinite monochromatic set, hence it is either constant or monotonic.

Question 15. Show that

$$\lim_{n \to \infty} \sum_{k=1}^{n} \frac{\sin(a^n)}{ka^n} = 0, \quad a \in \mathbb{R} \setminus [-1, 1]$$

Solution.   The elements of this sequence are all the partial sums. According to the Cauchy's criterion of the convergence, it is enough to show for arbitrary $n, m \in \mathbb{N} : |x_n - x_{n+m}| < \varepsilon$. We have

$$|x_n - x_{n+p}| = \left| \sum_{k=n+1}^{n+p} \frac{\sin(a^n)}{ka^n} \right| \leq \sum_{k=n+1}^{n+p} \left| \frac{\sin(a^n)}{ka^n} \right|$$

$$\leq \sum_{k=n+1}^{\infty} \left| \frac{\sin(a^n)}{ka^n} \right| \leq \sum_{k=n+1}^{n+p} \left| \frac{1}{ka^n} \right|$$

Since the last series is a convergent geometric series, it follows

$$|x_n - x_{n+p}| < \varepsilon.$$

Question 16. The following question explores the sequence

$$a_n = \left(1 + \frac{1}{n} \right)^n$$

1. Prove that the sequence a_n is increasing.

2. Prove that $2 < a_n < 3$ for all $n \in \mathbb{N}$.

3. Prove the inequalities:

$$m < n \Rightarrow \frac{1}{m^k} \binom{m}{k} < \frac{1}{n^k} \binom{n}{k} < \frac{1}{2^{k-1}}, \quad (m, n, k \in \mathbb{N})$$

4. Prove that $\left(1 + \dfrac{1}{n} \right)^n < \left(1 + \dfrac{1}{n+1} \right)^{n+1}$.

5. Show that $\lim_{n \to \infty} \left(1 + \dfrac{1}{n} \right)^n = e$. Use this sequence to define e^{-1}.

6. Show that

$$\lim_{n \to \infty} \sum_{k=0}^{n} \frac{1}{k!} = e$$

7. Prove that e is irrational.

Solution.

1.

$$\frac{a_{n+1}}{a_n} = \frac{\left(1+\dfrac{1}{n+1}\right)^{n+1}}{\left(1+\dfrac{1}{n}\right)^n} = \frac{\left(\dfrac{n+2}{n+1}\right)^{n+1}}{\left(\dfrac{n+1}{n}\right)^n}$$

$$= \frac{n+2}{n+1}\left(\frac{n(n+2)}{(n+1)^2}\right)^n = \frac{n+2}{n+1}\left(1-\frac{1}{(n+1)^2}\right)^n$$

Now using Bernoulli's inequality, we have:

$$\left(1-\frac{1}{(n+1)^2}\right)^n \geq 1 - \frac{n}{(n+1)^2} = \frac{n^2+n+1}{n^2+2n+1}$$

and together with previous identity we obtain:

$$\frac{a_{n+1}}{a_n} \geq \frac{n+2}{n+1}\frac{n^2+n+1}{n^2+2n+1} = \frac{n^3+3n^2+3n+2}{n^3+3n^2+3n+1} > 1$$

2. The first inequality follows from part 1. and the fact that $a_1 = 2$. The second inequality is proven using the binomial theorem:

$$\left(1+\frac{1}{n}\right)^n = 1 + \frac{n}{n} + \frac{n(n-1)}{2!}\frac{1}{n^2} + \cdots$$

$$+ \frac{n(n-1)(n-2)\cdots(n-n+1)}{n!}\frac{1}{n^n}$$

$$< 1 + 1 + \frac{1}{2!}\left(1-\frac{1}{n}\right)$$

$$+ \cdots + \frac{1}{n!}\left(1-\frac{1}{n}\right)\left(1-\frac{2}{n}\right)\cdots\left(1-\frac{1n-1}{n}\right)$$

$$< 1 + 1 + \frac{1}{2!} + \frac{1}{3!} + \cdots + \frac{1}{n!}$$

$$< 1 + 1 + \frac{1}{2} + \frac{1}{2^2} + \cdots + \frac{1}{2^{n-1}}$$

$$= 1 + \frac{1}{1-\dfrac{1}{2}} = 3$$

3. The statement follows from the binomial theorem applied on the sequences $\left(1+\dfrac{1}{n}\right)^n, \left(1+\dfrac{1}{m}\right)^m$ and knowing that $m < n \Rightarrow m^k < n^k$. Another way to prove is using part 1.

4. This also follows from part 1.

5. The sequence is convergent as increasing and bounded from above. The supremum of the sequence is *defined* to be the number e. Furthermore, we have:

$$\left(1-\frac{1}{n}\right)^n = \frac{1}{\left(1+\dfrac{1}{(-n)}\right)^{-n}} \to \frac{1}{e}, n \to \infty$$

6. We will prove:

$$\lim_{n\to\infty}\left(1+\frac{1}{1!}+\frac{1}{2!}+\frac{1}{3!}+\cdots+\frac{1}{n!}\right) = e$$

It is using the binomial theorem:

$$a_n = \left(1+\frac{1}{n}\right)^n = 1 + \frac{n}{n} + \frac{n(n-1)}{2!}\frac{1}{n^2} + \cdots +$$
$$+ \frac{n(n-1)\cdots(n-k+1)}{k!}\frac{1}{n^k} +$$
$$+ \frac{n(n-1)\cdots 2\cdot 1}{n!}\frac{1}{n^n} +$$
$$> 2 + \frac{1}{2!}\left(1-\frac{1}{n}\right) + \frac{1}{3!}\left(1-\frac{1}{n}\right)\left(1-\frac{2}{n}\right) + \cdots +$$
$$+ \frac{1}{k}\left(1-\frac{1}{n}\right)\cdots\left(1-\frac{k-1}{n}\right)$$

For any $k \in \mathbb{N}$ when $n \to \infty$: $e \geq \left(1+\dfrac{1}{1!}+\dfrac{1}{2!}+\dfrac{1}{3!}+\cdots+\dfrac{1}{k!}\right)$. However the inequality is strict, since the set of values of

$$\left(1+\frac{1}{1!}+\frac{1}{2!}+\frac{1}{3!}+\cdots+\frac{1}{n!}\right)$$

doesn't have the maximum element. Using proof of 2. above, we immediately have: $a_n < \left(1+\dfrac{1}{1!}+\dfrac{1}{2!}+\dfrac{1}{3!}+\cdots+\dfrac{1}{n!}\right), n \in \mathbb{N}$. This gives the desired limit.

7. Suppose $e = \dfrac{h}{k}, k > 1$. Since $e = \sum_{k \geq 0} \dfrac{1}{k!}$ we see:

$$c := k! \left(e - \sum_{i=0}^{k} \dfrac{1}{i!} \right) \in \mathbb{Z}$$

and replacing e with it's infinite series, we obtain

$$c = \sum_{i=1}^{\infty} \dfrac{1}{\prod_{j=1}^{i}(k+j)} \leq \sum_{i=1}^{\infty} \dfrac{1}{(k+i)^i} = \dfrac{1}{k} < 1$$

so that $c \in \mathbb{Z}, c < 1$ - contradiction.

Question 17. Prove that:

1. $\sqrt[n]{a} \to 1, n \to \infty, a > 0$

2. $\sqrt[n]{n} \to 1, n \to \infty$

Solution.

1. If $a = 1$, there is nothing to prove. Now let $a > 1$. Then we have

$$a = (1 + (\sqrt[n]{a} - 1))^n > n(\sqrt[n]{a} - 1)$$

and so

$$0 < \sqrt[n]{a} - 1 < \dfrac{a}{n} < \varepsilon$$

for all $n > \dfrac{a}{\varepsilon}$. If $0 < a < 1$, we take $b = \dfrac{1}{a} > 1$ and from previous case $\sqrt[n]{b} \to 1$. Then

$$\lim_{n \to \infty} \sqrt[n]{a} = \lim_{n \to \infty} \dfrac{1}{\sqrt[n]{\dfrac{1}{a}}} = \dfrac{1}{\lim_{n \to \infty} \sqrt[n]{a}} = 1$$

2. We have using binomial theorem:

$$n = \left(\sqrt[n]{n} - 1 \right)^n = 1 + n(\sqrt[n]{n} - 1) +$$
$$+ \dfrac{n(n-1)}{2}(\sqrt[n]{n} - 1)^2 + \cdots + (\sqrt[n]{n} - 1)^n$$
$$> \dfrac{n(n-1)}{2}(\sqrt[n]{n} - 1)^2$$

so that

$$\left| \sqrt[n]{n} - 1 \right| < \sqrt{\frac{1}{n-1}} < \varepsilon$$

for all $n > 1 + \dfrac{2}{\varepsilon}$.

Question 18. Show that

$$n(a^{1/n} - 1) \to \ln a, a > 0$$

Solution. From the inequalities:

$$\left(1 + \frac{1}{n} \right)^n < e < \left(1 + \frac{1}{n+1} \right)^n$$

taking logarithm to both sides:

$$1 < n(e^{1/n} - 1) < 1 + \frac{1}{n+1}$$

So:

$$\lim_{n \to \infty} n(e^{1/n} - 1) = 1$$

Let $a > 1$ and let

$$y_n = n(a^{1/n} - 1) = n(e^{\ln a/n} - 1) = \frac{n}{\ln a}(e^{\ln a/n} - 1)\ln a$$

Regarding $\dfrac{n}{\ln a}$, we will analyze:

$$\alpha_n = \left\lceil \frac{n}{\ln a} \right\rceil$$

and obtain:

$$\alpha_n \le \left\lceil \frac{n}{\ln a} \right\rceil \le \alpha_n + 1$$

and

$$\frac{1}{\alpha_n + 1} \le \left\lceil \frac{\ln a}{n} \right\rceil \le \frac{1}{\alpha_n}$$

Now we have:

$$\ln a\alpha_n(e^{1/\alpha_{n-1}-1}) < n(a^{1/n}-1) < \ln a\alpha_{n+1}(e^{1/\alpha_{n-1}}-1)$$

and

$$-\ln a(e^{1/\alpha_{n+1}}-1) + \ln a(\alpha_{n+1}(e^{1/\alpha_{n+1}}-1) < y_n$$
$$< \ln a\alpha_n(e^{1/\alpha_n-1} + \ln a(a^{1/\alpha_n}-1)$$

Now $\alpha_n(e^{1/\alpha_n-1})$ is a subsequence of $n(e^{1/n-1})$ which is convergent to 1.

When $0 < a < 1$, we have

$$y_n = n(a^{1/n}-1) = n(\frac{1}{a^{-1/n}}-1)$$
$$= \frac{n(1-b^{1/n})}{b^{1/n}} = -b^{1/n}n(b^{1/n}-1) \to -\ln b = \ln a, n \to \infty$$

Question 19. Find

$$\lim_{n\to\infty} \sum_{k=1}^{n} \frac{1}{n+k}$$

Solution. Two steps are required.

1.

$$\sum_{k=1}^{n} \frac{1}{n+k} = \frac{n}{2n+1} + \sum_{k=1}^{n} \frac{1}{(2k)^3 - 2k}$$

Proof. It is:

$$\frac{1}{x^3 - x} = \frac{1}{2(x+1)} + \frac{1}{2(x-1)} - \frac{1}{x}$$

Let $x = 2k$, sum over k, $a \le k \le n$. The right side from above is:

$$\frac{1}{2}\sum_{k=1}^{n}\frac{1}{2k-1} + \sum_{k=1}^{n}\frac{1}{2k+1} - \frac{1}{2}\sum_{k=1}^{n}\frac{1}{k} + \frac{n}{2n+1} =$$

$$= \sum_{k=1}^{n}\frac{1}{2k-1} - \frac{1}{2}\sum_{1}^{k}\frac{1}{k}$$

$$= \sum_{k=1}^{2n}\frac{1}{k} - \sum_{k=1}^{n}\frac{1}{k}$$

$$= \sum_{k=1}^{n}\frac{1}{n+k}$$

2. We will show that:

$$\lim_{n\to\infty}\sum_{k=1}^{n}\frac{1}{k} - \ln(n) = C$$

where C is Euler's constant.

Proof. From $x_{n+1}-x_n = \dfrac{1}{n+1} - \ln\left(1+\dfrac{1}{n}\right) < 0$ follows that the sequence is decreasing. Now we show that it is bounded from below.

$$\sum_{k=1}^{n}\frac{1}{k} - \ln n > \ln 2 + \ln\left(1+\frac{1}{2}\right) + \cdots + \ln\left(1+\frac{1}{n}\right)$$

$$= \ln\left(2\cdot\frac{3}{2}\cdot\frac{4}{3}\cdots\frac{n+1}{n}\cdot\frac{1}{n}\right) = \ln\left(1+\frac{1}{n}\right) > 0$$

Therefore, the desired limit is the highest lower bound (infimum) which is denoted by C.

With regard to the posed problem, we have:

$$\lim_{n\to\infty}(z_{2n} - z_n) = \lim_{n\to\infty}\sum_{k=1}^{n}\frac{1}{n+k} = \ln(2n) - \ln(n) + C - C = \ln 2$$

where

$$z_n := \sum_{k=1}^{n}\frac{1}{k}$$

Question 20. Find the limit

$$\lim_{n\to\infty} \frac{1}{n}\left(\left(x_0 + \frac{c}{n}\right)^2 + \left(x_0 + \frac{2c}{n}\right)^2 + \cdots + \left(x_0 + \frac{(n-1)c}{c}\right)^2\right)$$

Solution. Expanding all squares and removing the brackets and then factoring c and x_0 we obtain that the expression is equivalent to:

$$\frac{1}{n}\left(nx_0^2 + \frac{2cx_0}{n}\sum_{k=1}^{n}k + \frac{c^2}{n^2}\sum_{k=1}^{n}k^2\right)$$

$$= \frac{1}{n}\left(nx_0^2 + \frac{2cx_0}{n}\frac{n(n-1)}{2} + \frac{a^2}{n^2}\frac{(n-1)n(2n-1)}{6}\right) \to x_0^2 + cx_0 + \frac{c^2}{3}$$

Question 21. Show that

$$\lim_{n\to\infty}(\sin\sqrt{n+1} - \sin\sqrt{n}) = 0$$

Solution. We have:

$$\left|\sin\sqrt{n+1} - \sin\sqrt{n}\right| = \left|2\sin\frac{\sqrt{n+1}-n}{2}\cos\frac{\sqrt{n+1}+\sqrt{n}}{2}\right|$$

$$\leq 2\left|\sin\frac{1}{2(\sqrt{n+1}+n)}\right| < \frac{1}{\sqrt{n+1}+n} < \frac{1}{2\sqrt{n}} < \varepsilon$$

whenever $n > \left\lceil\dfrac{1}{4\varepsilon^2}\right\rceil$.

Question 22. Find

$$\lim_{n\to\infty}\left(\frac{n+3}{4n+1}\right)^{n^3+2n+1}$$

Solution. If $f(n) = \dfrac{n+3}{4n+1}$, $g(n) = n^3 + 2n + 1$, we analyze the function

$$e^{g(n)\ln f(n)}$$

that is:

$$e^{(n^3+2n+1)\ln\left(\frac{n+3}{4n+1}\right)}$$

Since for n large enough: $n + 34n + 1 < 1$, then $\ln\left(\dfrac{n+3}{4n+1}\right) < 0$ and $g(n) > 0$ it follows that the desired limit is zero.

Question 23. Depending on parameters x and a find the limits of the sequence

$$\lim_{n\to\infty} \sqrt[n]{1 + \left(\frac{x}{a}\right)^n}, \quad x > 0, a > 0$$

Solution. Two cases must be distinguished:

Case 1. $0 < \dfrac{x}{a} \leq 1$, that is $0 < x \leq a$, and in this case the inequalities:

$$1 \leq \sqrt[n]{1 + \left(\frac{x}{a}\right)^n} \leq \sqrt[n]{2} \leq 1$$ hence the limit is 1.

Case 2. For $x > a > 1$, we have:

$$\sqrt[n]{1 + \left(\frac{x}{a}\right)} = \frac{x}{a} \sqrt[n]{1 + \left(\frac{a}{x}\right)^n} \to \frac{x}{a}, n \to \infty$$

Therefore:

$$\lim_{n\to\infty} \sqrt[n]{1 + \left(\frac{x}{a}\right)^n} = \begin{cases} 1, & \text{if } 0 < x \leq a \\ \dfrac{x}{a}, & \text{if } a < x < \infty \end{cases}$$

References

1. Heuser, H. *Lehrbuch der Analysis*, Springer, 2009.

2. Zorich, V. *Mathematical Analysis 1*, Springer, 2015.

3. Khinchin, A. Y., *A Course in Mathematical Analysis*, Hindustan Publish Corp., 1960

4. Smirnov, V. I., *A Course of Higher Mathematicsm Vol I*, Pergamon Press, 1964.

5. Berndt, B. C., *Ramanujan's Lost Notebooks I*, Springer, 1985.

Contents

www.ingramcontent.com/pod-product-compliance
Lightning Source LLC
Chambersburg PA
CBHW050759290526
45792CB00008B/2252